Two Centuries of Telecommunications in My Lifetime

by Bill Rosoman

ISBN 978-1-927157-28-2

Table of Contents

About the Author

I have lived an interesting and varied life. Currently I am unemployed and living in a Mobile Home (RV) mostly at a property I have purchased out the back country. Life is full of surprises and chances.

I really regret little in my life, of cause if I had my life over again there are things that I would change. But hey life goes on and follows it's path.

I am currently publishing and writing some more ebooks. This is interesting and fun. I am also making some money!

I have worked with technology for many years and have been quite prophetic over the years. I have been working on Books and Ebooks online for many years now. It is starting to pay some dividends and we are starting to get some rewards.

I am a Pacifist at heart but if push comes to shove would defend myself. If my Life was Threatened the I Would Act.

I consider myself an Optimistic Realist.

"Optimism has a Chance

Persimmon has no Chance."

Life has had its ups and downs, but general my life has been and is great.

My glass is always half full.

Disclaimer

I have taken care to identify people and web pages I have accessed or used. I acknowledge all copyrights and trademarks.

Other websites;

Personal Site

www.webng.com/leftfieldnz

Photos:

http://picasaweb.google.com/leftfieldnz

Movies:

www.youtube.com/leftfieldnz

Business Site

www.creativekiwis.com

Blog

http://leftfieldnz.wordpress.com

Telecommunication In My Lifetime

Were do I start, in my lifetime things in the telecommunications department have changed so much it is not funny.

New Zealand Telecommunications Timeline

http://www.wordworx.co.nz/KiwitelcoTimeline.htm

This is the brief history of Telecommunications in New Zealand; http://www.teara.govt.nz/en/rural-services/5

Communications

Early postal services

The New Zealand Post Office was set up in the 1840s, but until the 1860s services were infrequent because poor roads made it difficult to carry mail between settlements.

Horses and horse-drawn coaches transported mail until the 1920s. Pigeon post was used in some rural regions. From the 1860s the number of post offices began to increase, and by 1900 there were around 1,700 branches throughout the country.

Popular postie

In 1868 William Baines had the contract to carry mail fortnightly between Christchurch and Timaru. A small man with a neat beard and a limitless fund of gossip, he was a welcome caller at homesteads on his route, as he delivered not just mail but also the latest news.

Rural delivery

Before 1905 country people had to collect their mail from the nearest post office, but when a rural delivery service began, farmers could both receive and send mail through a special rural delivery post box or bag at the farm gate. Rural delivery was extended in the 1920s, when motorcycles were used to carry the mail. This was a daily service in some places, just once or twice a week in others. Essentials such as newspapers, groceries and animal feed were also dropped off.

Mail and other goods were often delivered on the school bus as it took children to and from the district school. Correspondence school lessons and assignments were received and sent through the rural delivery service, which was still being operated by New Zealand Post in 2008.

Rural post offices

In small country towns, the post office had many functions aside from delivering mail and newspapers. People went there to receive pensions and allowances, open savings accounts, enrol to vote, and pay car registrations and other fees. The postmaster was often the registrar of births, deaths and marriages, a witness for statutory declarations and an adviser on government services.

There was strong protest from rural dwellers when many country post offices were closed between 1981 and 1991. This occurred because of economic changes – a farming downturn led to depopulation of country towns, and the government restructured the Post Office and sold its banking and telephone functions.

Telecommunications

Telegraph

Electric telegraph messaging using Morse code was introduced to New Zealand in 1862, connecting provincial towns with ports. In 1863 the Electric Telegraph Department began; it merged with the Post Office in 1881. Telegraph lines were erected throughout the country, and by 1891 extended to many small rural towns.

Radio

Wireless telegraphy – sending messages by radio – was seen as a possible means of communication for rural people in the early 20th century. However, under the Wireless Telegraphy Act 1903, only the government could receive and transmit radio messages. It was illegal for citizens to do this until 1921, by which time the telephone was becoming established. Radio telephones were later used on some farms – in the late 1940s, a system linked homesteads along the

Rakaia River in Canterbury with the Methven post office.

Telephone innovation

The first telephone line into the Conway Flat area in North Canterbury was built in the 1930s. The line was insulated using glass bottles with their bottoms knocked out, which were jammed into mānuka poles.

Telephone

The telephone was introduced in 1881, but was not widely used until after 1890. Some country districts had telephones earlier than others; there were lines in rural Canterbury by 1904. Many farmers erected their own lines before the Post and Telegraph Department reached outlying areas.

The telephone became essential for running a farm business. It allowed farmers to check prices, order goods and contact neighbours. It also saved time; for instance, farmers could ring the railway station instead of making fruitless trips to check for incoming goods. The phone was a social lifeline, especially for farmers' wives.

After the Second World War, small automatic telephone exchanges were set up in rural areas, mostly with small party (shared) lines with up to 10 customers. Because it was possible to listen to other people's conversations, party lines were a rich source of local gossip. They were common until the 1960s, but few remained in the early 2000s.

Subsidising rural services

In 1987 the telecommunications division of the New Zealand Post Office became a corporation, Telecom. In 1990 it was privatised, and began competing with other telecommunications companies. The Telecommunications Service Obligations (TSO) contract between Telecom and government ensured that rural telephone services were subsidised, but in 2008 some country areas still did not have adequate land-line or cellular services.

Also reminds me of my nephew Haven when he was about ten, I said to him, "boy in my day there was no calculators aloud in an exam, no TV, no cell phones, no computers, no DVD players, no radio telephones." He looked at me and said, "bullshit Uncle". He had no conception of not having these modern contraptions.

Fax

Fax (short for facsimile), sometimes called telecopying.

Fax machines took over in large part Telex Machines which died in New Zealand in the 1980s.

Before the internet they were a great way to get information sent quickly and at the time cheaply.

However with the advent of Email and then the Internet, Faxes have died away, although still available in 2012,

Telex

Telex Machines were used by large Business Organisations in the 1960-80 era.

Telex was the forerunner of modern email and texting

They were used sometimes like sending a Telegram say for Motel Bookings, and sometimes sending large amounts of data using a ticker tape to speed up the input.

Telegram

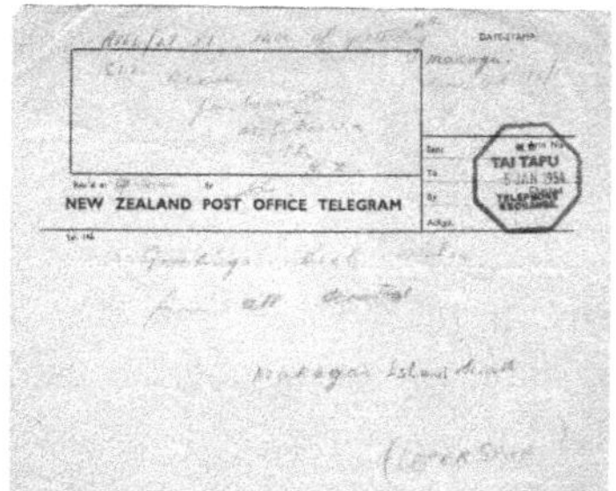

Sending Telegrams was the most common way of communicating for many years.

As I worked for the Post Office I got free Telegrams so that was great.

Many people did not have a land line, cell phones had not been invented yet and sending a Telegram was a cheap way of getting a message through to someone usually in a matter of hours.

Even if people did have a Telephone it was quite a mission to make a call and it was expensive.

I see you can still send a Telegram to 200 countries.

http://www.itelegram.com/

Pacnet

1983: United Nations World Communications Year, celebrating the imminent convergence of computers and communications technology. The New Zealand Post Office began trialling a satellite-based X.25 packet switching data transmission service (Pacnet) and installed the first fibre optic cable in a 14km pilot system between Wellington and Lower Hutt carrying several hundred telephone circuits.

Pacnet was great, we used it in the early eighties to connect to databases like the National Library or Mcdonnell Douglas Corporation super computer in the USA.

We had to have a computer, a modem, a telephone line, a NZ Telecom Pacnet account and access to the database we wished to use.

I remember it was costing me round $200 per month for Pacnet and when we dialled into the USA it was 30 digits of a number.

I think we used to use 08743 at 2400 baud rate.

This is some information I got off the Internet.

PACNET AND KEEPING IN TOUCH

Information on connection

To help with phreaking into the Pacnet system, the following information is offered for your use

Pacnet is a network that lets you talk to online systems around New Zealand and worldwide.

DIALING PACNET:

To call Pacnet in your local area, dial:

1740 or 08740 for 300 baud

1741 or 08741 for 1200/75 baud

1742 or 08742 for 1200 baud

1743 or 08743 for 2400 baud <-- apparently this is secret!!

When you connect to the Pacnet system, you will get a message like this...

PACNET 002 005 057 021

CONNECTING TO OTHER SYSTEMS:

The formats to request a connection are:

R(NUA) [Return]Example: ?R-4700028 [Return]

?N(NUI)-(NUA) [Return] Example: ?N(Password)-4700048 [Return]

Where: NUI = Network User Identity

(your magic password)

NUA = Network User Address

(the address of who you want to call)

The first example requests a reverse charge call to Parliament Print in Wellington, the second requests connection to the National Library Computer System also in Wellington.

In the second example, a 'magic password' is needed (one of which will be given later).

Provided that the remote system accepts reverse charging, or your password is valid, you will get a message indicating that you are connected from Pacnet -

COM <-- this means you are COMnected!!

NOTES ON CHARGING:

There is no charge to your phone for calls unless you use a password that you apply for. Everything gets billed to the called party, or password owner.

NOTES ON THE NUI FORMAT:

All passwords that have been see so far appear to be eight (8) characters in length. The initial part is supplied by the customer, the remaining are numeric digits which are added by Telecom to pad the password to eight characters

PaSsWoRdS are case sensitive!!

Examples: Hello123 zZ123456 41587324

Note that passwords do NOT echo back to you. For example, if you typed in

"?NHello123-30000024", all you would see on your terminal would be "?-30000024".

A valid password at the time of the writing of this document is "a3035032".

Use this to your heart's content!!

NOTES ON THE NUA FORMAT:

An address is eight digits (usually) for local calls.

Example: 55999999

Where: 55 is the area code

999999 is the number of the party to be called

in that area

Other known area codes are: 30 -- Christchurch

47 -- Wellington

97 -- Auckland

For international calls, the format is slightly different.

Example: 01111999...9

Where: 0 is the indication of being international

1111 is the country code

999...9 is the remote network address you

wish to call

New Zealand has a country code which is 5301, but you never need to type this in when calling locally.

NUA's THAT ACCEPT REVERSE CHARGING:

Here are a few addresses that accept reverse charging:

(5301) 30000015 --- [COM] PRIME-NET
(5301) 30000047 --- [COM] LINCOLN UNIVERSITY
(5301) 47000028 --- [COM] GOVERNMENT PRINT
(5301) 47000045 --- [COM] LCG??
(5301) 97000035 --- [COM] BNZSR - PrimeNet
(5301) 97000038 --- [COM] Type "Hello" "//ONxxxxxxxx"

To try these, dial up Pacnet as described earlier, then type "?R-(NUA) [Ret]"

NUA's THAT CAN BE USED WITH THE PASSWORD GIVEN:

Here are a few addresses that can be used with the password given earlier:

045012201111 Get you to where you can get a hold of olypmic results,

messages, altlete profiles etc.

24000016XMicro public BBS run by Otago University (free). Type

XMICRO at the vax login prompt.

026245300040023 Technical University of Berlin (online chats)

47000048National Library Computer System

023421920100515 Hostess, UK

023421920100605 GMT Check, UK

0310690157800 BIX

30000034Canterbury University

30000047Lincoln University

30000085CHMEDS

4600800060 WNV.DSIR.GOVT.NZ

47000000WNV.DSIR.GOVT.NZ

4700000203 - RANZFS

4700000304 - RBNZFS

4700000405 - RCNZFS

47000013MAF Wellington, 04 MAF.GOVT.NZ

47000034NZBN

47000049Victoria University

47000090WCC.GOVT.NZ

48000046WNMEDS

97000073Auckland University

To try these, dial up Pacnet, then type "?Na3035032-(NUA) [Return]"

Thanks to all the people who helped me to compile this document.

[BROUGHT TO AUCKLAND, NEW ZEALAND COURTESY OF THE BANANA REPUBLIC BBS]

by Arnold Schwarzenegger

Starnet

Starnet was a pre internet service, offering mostly email to 25 mainly OECD countries. My email for Starnet was rde002.

It was handy to contact people quickly though it was expensive.

Bulletin Board System

From Wikipedia, the free encyclopedia

A Bulletin Board System, or BBS, is an online service based on microcomputers running appropriate software. Once logged in, users can upload and download software and data, read news and bulletins, and exchange messages with other users either through email or in public message boards. Many BBSes also offer on-line games, in which users can compete with each other, and BBSes with multiple phone lines often provide chat rooms, allowing users to interact with each other more instantaneously.

Ward Christensen coined the term "Bulletin Board System" as a reference to the traditional cork-and-pin bulletin board often found in entrances of supermarkets, schools, libraries or other public areas where people can post messages, advertisements, or community news. By "computerizing" this method of communications, the name of the first BBS system was born: CBBS - Computerized Bulletin Board System. During their heyday from the late 1970s to the mid-1990s, most BBSes were run as a hobby free of charge by the system operator (or "SysOp"), while other BBSes charged their users a subscription fee for access, or were operated by a business as a means of supporting their customers. Bulletin board systems were in many ways a precursor to the modern form of the World Wide Web, social network services and other aspects of the Internet.

As the use of the Internet became more widespread in the mid to late 1990s, traditional BBSes rapidly faded in popularity. Today, Internet forums occupy much of the same social and technological space as BBSes did, and the term BBS is often used to refer to any online forum or message board. Although BBSing survives only as a niche hobby in most parts of the world, it is still an extremely popular form of communication for Taiwanese youth (see PTT Bulletin Board System). Most BBSes are now accessible over telnet and typically offer free email accounts, FTP services, IRC and all of the protocols commonly used on the Internet. Some offer access though packet switched networks, or packet radio connections.

Early BBSes were often a local phenomenon, as one had to dial into a BBS with a phone line and would have to pay additional long distance charges for a BBS out of the local calling area. Thus, many users of a given BBS usually lived in the same area, and activities such as BBS Meets or Get Togethers were common, where users of the board would gather at a local restaurant, the SysOp's home or similar venue and meet face to face.

These were great though again expensive. You could upload or download files. You could send an email, though they bounced around the planet and were not instant like today.

You would send an email via your local BBS service, though would log onto another BBS say in Auckland once or twice a day, that BBS would log onto say Sydney who would log onto LA etc. and the same would happen in return. So it may have taken 24 hours or more for an email to bounce there and back again.

Technical Stuff

I have learnt a bit about electronics over the years.

I use this formula wheel all the time.

For those of you with mathematical brains
here is a formula chart to help you.

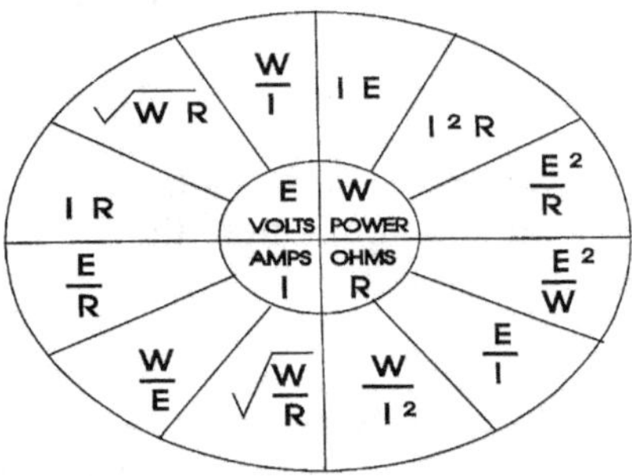

BTW they used to use V for volts now they use E for volts. E is for EMF (from Wikipedia, Electromotive "force" is not a force (measured in newtons) but a potential, or energy per unit of charge, measured in volts).

Other things we used in my Telecom days were

Why Ring Bells You People

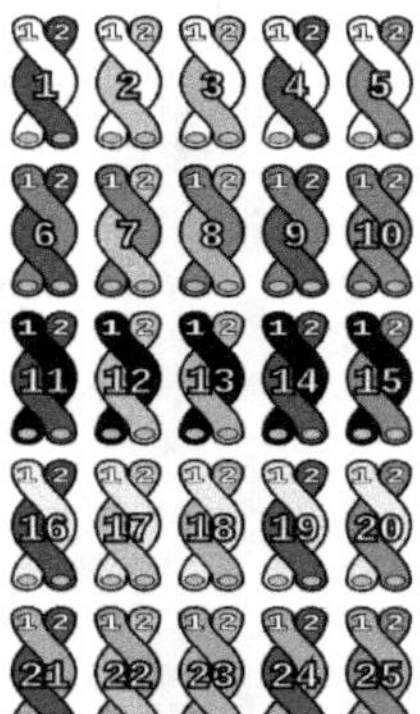

This is a saying we used to remember the Telephone Cable Colour Code for plastic cables.

Basically you had 5 colours and their combinations.

White, Red, Black, Yellow, Purple are the colours.

Sometimes they were just the colours or they may use stripped colours as well. And they used colour wraps for groups of 25 pair and then 100 pair etc.

This is similar to the cables I was used to using.

There was also lead covered cables were the wires were covered in paper using basic colours and strips. These were up to 2000 pairs of fine wires.

Did a bit of Cable Pulling in my time.

Victory Over Red Indians

This is a saying we used to remember Ohms Law.

http://en.wikipedia.org/wiki/Ohm's_law

Ohm's law

From Wikipedia, the free encyclopedia

This article is about the law related to electricity. For other uses, see

Ohm's acoustic law.

V, I, and R, the parameters of Ohm's law.

Ohm's law states that the current through a conductor between two points is directly proportional to the potential difference across the two points. Introducing the constant of proportionality, the resistance,[1] one arrives at the usual mathematical equation that describes this relationship:[2]

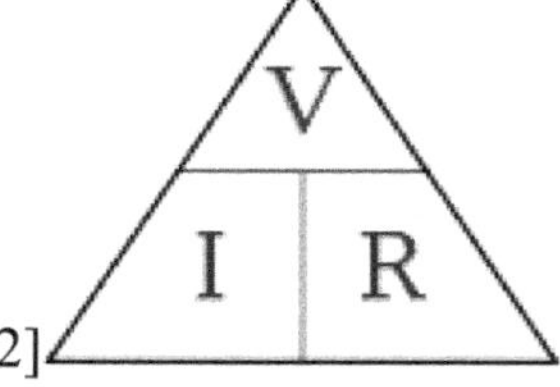

where I is the current through the conductor in units of amperes, V is the potential difference measured across the conductor in units of volts, and R is the resistance of the conductor in units of ohms. More specifically, Ohm's law states that the R in this relation is constant, independent of the current.[3]

The law was named after the German physicist Georg Ohm, who, in a treatise published in 1827, described measurements of applied voltage and current through simple electrical circuits containing various lengths of wire. He presented a slightly more complex equation than the one above (see History section below) to explain his experimental results. The above equation is the modern form of Ohm's law.

Pythagorean theorem

http://en.wikipedia.org/wiki/Pythagoras#Pythagorean_theorem

Since the fourth century AD, Pythagoras has commonly been given credit for discovering the Pythagorean theorem, a theorem in geometry that states that in a right-angled triangle the area of the square on the hypotenuse (the side opposite the right angle) is equal to the sum of the areas of the squares of the other two sides— that is,

.

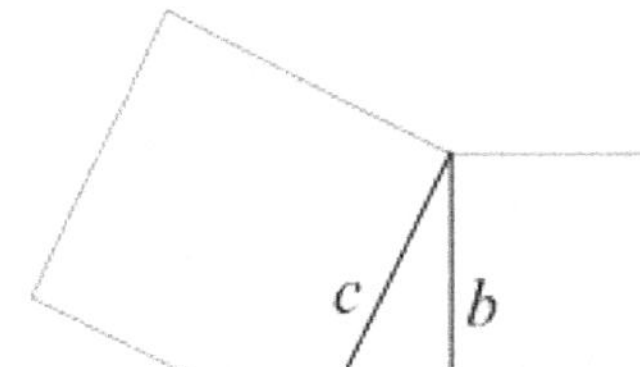

While the theorem that now bears his name was known and previously utilized by the Babylonians and Indians, he, or his students, are often said to have constructed the first proof. It must, however, be stressed that the way in which the Babylonians handled Pythagorean numbers implies that they knew that the principle was generally applicable, and knew some kind of proof, which has not yet been found in the (still largely unpublished) cuneiform sources.[46] Because of the secretive nature of his school and the custom of its students to attribute everything to their teacher, there is no evidence that Pythagoras himself worked on or proved this theorem. For that matter, there is no evidence that he worked on any mathematical or meta-mathematical problems. Some attribute it as a carefully constructed myth by followers of Plato over two centuries after the death of Pythagoras, mainly to bolster the case for Platonic meta-physics, which resonate well with the ideas they attributed to Pythagoras. This attribution has stuck down the centuries up to modern times.[47] The earliest known mention of Pythagoras's name in connection with the theorem occurred five centuries after his death, in the writings of Cicero and Plutarch.

I actually got to use Pythagoras when we erected a 75 foot Radio Mast onto Mount Hikurangi Spur in Ruatoria.

We used a 3-4-5 triangle to get the stays holding the mast at the right angle.

This is how we raised a large mast, by using a jury pole. (photo taken at Christchurch Rigging School in 1980s)

So some of the stuff I learnt at High School came in handy.

Some of the old Telephones we had

Manual (windup) Phones

Dial Phones (in the early days we only had black phones, there was a desk one and a wall one)

DTMF Phone.

Dual-tone multi-frequency signalling (DTMF) is used for telecommunication signalling over analog telephone lines in the voice-frequency band between telephone handsets and other communications devices and the switching centre. The version of DTMF that is used in push-button telephones for tone dialling is known as Touch-Tone. It was developed by Western Electric and first used by the Bell System in commerce, using that name as a registered trademark. DTMF is standardized by ITU-T Recommendation Q.23. It is also known in the UK as MF4.

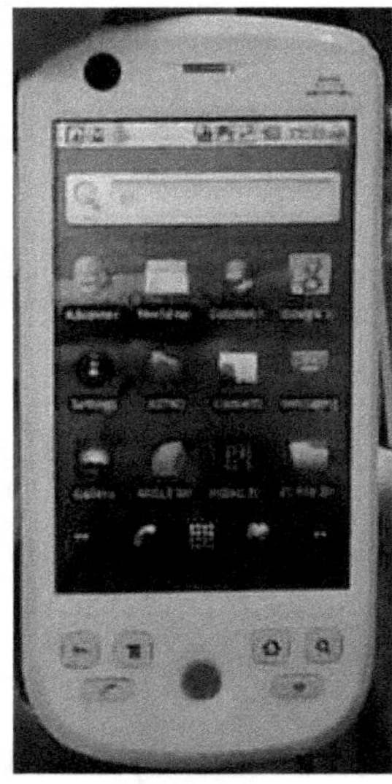

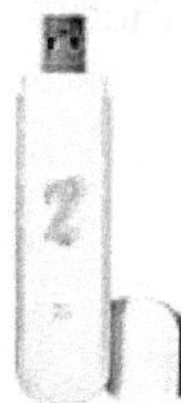

Smartphone (Android) and 3G USB Modem Stick

By 2013 there will be 4G in New Zealand and far better Internet Broadband available.

THE MORSE CODE

ALPHABET.

a ·—	j ·———	s ···
b —···	k —·—	t —
c —·—·	l ·—··	u ··—
ch ————	m ——	v ···—
d —··	n —·	w ·——
e ·	o ———	x —··—
f ··—·	p ·——·	y —·——
g ——·	q ——·—	z ——··
h ····	r ·—·	é ··—··
i ··		

FIGURES.

1 ·————	5 ·····	8 ———··
2 ··———	6 —····	9 ————·
3 ···——	7 ——···	0 —————
4 ····—		

In the Manual Telephone days the ring was probably in Morse code as most lines were a party line with many homes on the same line. Your phone number could have been 9A Tokomaru Bay and you would listen for a short and a long ring.

Crystal Radio

From Wikipedia, the free encyclopedia

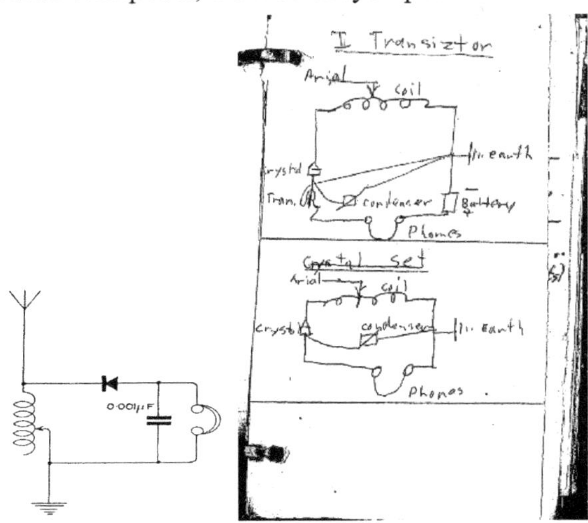

A crystal radio receiver, also called a crystal set or cat's whisker receiver, is a very simple radio receiver, popular in the early days of radio. It needs no battery or power source and runs on the power received from radio waves by a long wire antenna. It gets its name from its most important component, known as a crystal detector, originally made with a piece of crystalline mineral such as galena.[1] This component is now called a diode.

I made a crystal radio in 1958 aged 11.

The pic is the original from my diary I still have.

I used to hid under the blankets and listen to serials like The Scarlet Pimpernel etc.

Valve Radio 

We had one in the Lounge and I used to listen to overseas radio stations like The Voice of America, BBC, Radio Havana, Radio Peking. (see elsewhere for QSL cards)

The radio had medium wave for normal Broadcasting Stations and Short wave or Worldband.

http://en.wikipedia.org/wiki/Shortwave_bands

The bands and frequencies below are derived from multiple sources, and different radios may have different frequency numbers. Most international broadcasters use amplitude modulation with 5 kHz steps between channels; a few use single sideband modulation.[1] The World Radiocommunication Conference (WRC), organized under the auspices of the International Telecommunication Union, allocates bands for various services in periodic conferences. The last WRC took place in 2007. At WRC-97 in 1997, the following bands were allocated for international broadcasting:

Band	Frequency Range	Remarks
120 m	2300 - 2495 kHz	Tropical (regional) band
90 m	3200–3400 kHz	Tropical band
75 m	3900–4000 kHz	Shared with the North American amateur radio 80 m band
60 m	4750–5060 kHz	Tropical band
49 m	5900–6200 kHz	
41 m	7200–7450 kHz	Shared with the amateur radio 40 m band
31 m	9400–9900 kHz	Most heavily used band
25 m	11,600–12,100 kHz	
22 m	13,570–13,870 kHz	Substantially used in Eurasia
19 m	15,100–15,800 kHz	
16 m	17,480–17,900 kHz	
15 m	18,900–19,020 kHz	Lightly utilized; may become DRM band in future
13 m	21,450–21,850 kHz	
11 m	25,600–26,100 kHz	May be used for local DRM broadcasting

My work with Post Office/Telecom

I started my career at Wanganui, my home town in 1965 as a Junior Lineman for the New Zealand Post Office. It was formally known as the P&T or the Post and Telegraph Department or the Puffed and Tired. I spent time also in Wellington on temporary transfer making lots of money but paying sixty cents in the dollar tax.. It was a time were you waited quite happily for three months to get the phone and then you had a choice of a black phone or nothing.

In 1970 I was promoted to Tokomaru Bay as a Foreman.

In 1974 Tokomaru went the Telephone Exchange Automatic and I moved to Ruatoria were the Telephone Exchange went Automatic in 1974 and were I stayed till being made redundant in 1988.

When I arrived in Tokomaru Bay in 1970, the phones were all manual and wind-up jobs. I was the Faultman or in charge of the two man Installing/maintenance gang.

We were responsible for all the outside plant of underground cables, telephone poles and wires and radio masts and aerials.

Old Manual Ruatoria Exchange 1976 (Mrs Keelan)

There was a Postmaster a Post Office/Post Bank staff of about four people, there was the Telephone Exchange manned 24 hours a day by about ten staff, some part time and the Line staff of a Senior Foreman and Line gang of four or five, an Installing gang of two and one Faultman. The Area in Tokomaru Bay was from south of Tolaga Bay, almost to Whangara and North to north of Waipiro Bay.

That is an area of some 4,000 square kilometres. We had no maps no radio just a lot of local knowledge.

New NEAX Automatic Exchange Ruatoria 1977

On two occasions in Tokomaru Bay they were thinking of organising search parties for me as I had not returned at night. I remember one clearly, I was working on a phone fault out the back of the Mata Road, which is about 40km from Tokomaru Bay on a gravel road, and most of that was open aerial lines on poles, so you had to isolate the fault to a section of line and then drive, walk and use aids like binoculars to locate the fault. If it was across country you had little choice but to carry a ladder and a kit of gear and get on with it. This time I decided that I was close to fixing the problem and it would be another big trip if I had to come back the next day. So I carried on and fixed the fault. They grilled me and said why did you not ring, I said the phone was out of order and I had no real way of contacting base to let them know.

In Ruatoria we had staff of around 14 at its height. There was an Overseer the last being Laurie Brown, I think one Senior Foreman being myself with the line gang and large line truck and hiab lifter. The line gang was usually Muss Kaika, Buck Tewhitu, Rewiti Green (till he died) Maurie Read (Tuwharetoa) Len Harkins, an Installing gang of two, two faultmen, one north of Ruatoria and one south, a storeman and cleaner Allan. Some of the other staff were Stan Hovell, Bill Blane, Harry Haenga, Kevin Bradley, Bremner Mill and some outsiders that used to come and go.

Ruatoria Line Staff + Gisborne Staff in 1970s

Back row Muss Kaika, Ray, ?, Len Harkins, Togi Waitoa, Dan Poihipi, ?

Front row Maurie Reed, Harry Haenga, Bert Moles, Buck Tewhitu, Max Davis, Solomon, helicopter Pilot

There was John Otton, Dan Poihipi, Winston Hamon as foremen over the years.

Then there was the manual telephone exchange and the Post Office down the road. That was around another 15 staff.

Telecom House
Cnr Dickens and Hastings Streets
Private Bag NAPIER

Telephone (070) 353 299 Ex 8512
Telex 3302 Fax (070) 351 165

31 August 1988

Telecom

TNA 104968
J Wilson

CERTIFICATE OF SERVICE

TO WHOM IT MAY CONCERN

This is to certify that

WILLIAM JOHN ROSOMAN

served in the New Zealand Post Office and the Telecom Corporation of New Zealand Ltd as stated hereunder

Commenced as a Junior Lineman Wanganui	18 January 1965
Promoted to Skilled Lineman Wanganui	22 October 1968
Promoted to Foreman Tokomaru Bay	20 July 1970
Transferred to Foreman Ruatoria	04 March 1974
Promoted to Senior Foreman	01 April 1976
Ceased due to Voluntary Severance	20 September 1988

Mr Rosoman rendered very good service.

This certificate is issued without alteration or erasure.

R S MATHER
District Manager

Telecom Corporation of New Zealand Limited

Telecom Corporation like the Post Office Government Department before it was not very efficient. Large corporations are inefficient by nature as far as I am concerned. They introduced a system were by a subscriber would have a telephone fault and the local postmaster or telephone exchange supervisor would send a Telegram to Gisborne who would send a Telegram to Ruatoria and then I would be notified. Whoever thought of this system was an idiot. I used to meet people in the street and they would say how come I had not fixed their fault and I would say I had not received it. Anyway I really peeved one day and rang I think Mac Odell the District Engineer and swore at him about the crazy system. It was risky for one you were supposed to use the chain of command and not ring the boss directly, let alone swear at him. But he said I understand and the silly system was changed.

You may think from this that it was all doom and gloom. But I did enjoy my time with Telecom as you felt you providing a public service which they mostly appreciated. We were also fairly well paid and our cost of living was not too bad. You were also in charge and out and about the country side, you had to be your own man and make some life decisions sometimes, if the weather was real bad the roads were shocking and even though you had a Landrover four wheel drive, it was sometimes risky out there and there was little backup if it turned to custard.

We had situations like an old Aunty would keep ringing the exchange because she wanted someone to talk too. They would say Bill go and see Aunty. She would be very happy to see and you would have a cup of tea and scones and she would be happy for awhile, The same can be said of the families like the Williams families who owned a lot of the big farms in the region. That goes back to Williams the Missionary who had a bible in one hand and took land in the other. The wives would welcome you and give you a cup of tea and all would be fine. The only exception was that if the Williams wives saw you in town or at a function they would not acknowledge you as they had to maintain their status in the community.

The generosity of the community came out at times like Christmas when farmers and business people as well as the public would drop off food, meat, chocolates and even beer to the exchanges for their appreciation of the service offered.

For quite a few years I was in charge of the rigging requirements on the East Cape. I had been to rigging school in Christchurch and found that very enjoyable. The first day there they asked you to climb a 50 foot pole, I said why the guy said do it. It was to sort out scared people who were sent home if they refused.

I was maintaining and installing mostly radio aerials and the links from the East Coast to the outside world via Microwave links.

We also used to maintain the aerials at the East Cape lighthouse were they had some HF radios.

We worked with the radio technicians from the Napier Radio depot. That was some 400+km from Ruatoria, so sometimes I would give the radio gear the drop test (drop it from say 25mm and see what happens). Or just unplug it and plug it back in, at least have a look and see what I could see. As it was a several day journey if the technicians had to come and fix things. We also worked in some bad conditions sometimes and did quite a bit of helicopter work. They also started putting in what they called country radio telephones as a way of replacing the many hundreds of kilometres of open aerial lines and telephone poles going into the back country. They were beamed from Hikurangi Spur or somewhere handy sometimes on mains power and sometimes on solar power. The radio link was in the middle of conventional telephone wiring and equipment.

We did carry some spares especially for the Country Radio sets. We would install the spares and then send the originals back to Napier for repair.

I used to get many allowances, like height allowance, rigging allowance, flying allowance it was great.

I can still splice ropes good, but nobody wants to know as nowadays the ropes are all synthetic not like the jute ones we used to use. Now you can just burn the ends and that is that.

Of cause all that has gone for good now as the farmers have mostly sold out to forestry and the large public service and corporations like Telecom are long gone. My friend John son used to do the East Coast telephone faults from Gisborne in the 1990s. He had to travel for hours and his wage was not that fantastic. But the big boys only worry about profits not service, efficiency or the workers.

In the early days we used to have Camp jobs. Where we would take a whole lot of poles, arms and wire and run overhead telephone lines in the back blocks, upgrading their telephones. Sometimes there would be one telephone line with say 5-6 houses and phone numbers on the same line. We would run an extra set of wires or maybe more. They had phone numbers like 3S which is - - - in Morse code or perhaps 9A, - . or the other way round. Everybody knew who was ringing who and what the

local gossip was,

We used to manhandle the poles up the hills on our shoulders and dig all the holes by hand. It was all manual work or if we were lucky we might borrow a horse from a local farmer.

Sometimes we would stay in a shearers quarters or perhaps and old house and we would take a mobile generator for power.

It was not till the 80s-90s that we started using helicopters for carting gear and personnel into the back and beyond.

The last camp job I remember was HoreHore Station at the back of Makarika, Ruatoria. We were living in the shearers quarters on the farm and set up camp for a week. Dennis flew us in and I said to him we need some meat. He was back in less than half an hour with two wild pigs for our week.

I think this is the shearers quarters in the photo above and it is the Radio Link in the photo right.

We put in a solar powered remote Radio Telephone link for the shearers quarters. They used a it couple of times a year for mustering or doing shearing and as an emergency phone (there was a person who drowned in the river on the way to the station). They used to pack out the wool in small bales on pack horse in the early days.

I guess the link was to Hikurangi Spur and the radio telecommunications base their. It was just on the other side of the mountain.

So this job was done in the mid 1980s

The station was owned by Les Brown at the time.

I remember camping out on the Ihungia Road, Waikura Valley as well HoreHore Station. I was also in charge of the Ruatoria Line Depot when Jack Barber a foreman from Gisborne was in camp at Lotton Point upgrading their telephone systems.

Mentioning Waikura Valley reminds of the funniest telephone fault we had in all the years of my working for Telecom.

At the shepherds house at Oweka Station, right at the end of the Waikura Valley, they had a rural Radio Telephone base and from there it went to the other places on the farm including the Manager/Owner Gordon McKay I think he was.

Anyway after awhile the phone was going on the blink all the time and I would drive the two hours each way by Landrover and by the time I got there the telephone would be fine. The shepherds wife would invite me in for a cup of tea and I would head home to Ruatoria. This went on for weeks. The District Engineer Murray got suspicious. I then flew in by helicopter with Dennis one day. It was not a nice day and we flew above the trees but below the clouds. The air trip took eleven minutes each way. But again by the time we got there it was all OK.

No one could figure out what was happening till one day the Manager shot up to the shepherds house and found the fuse pulled out of the radio so it would not work. It turned out the shepherds wife was lonely and had pulled the fuse, and would replace when she saw me coming up the drive. The shepherd was fired and all was well from then on. The overall cost was around $20,000 for my time, the vehicles and the helicopter.

The Owner of Oweka Station used to fly his private single engine plane and the family to Whakatane to go shopping. Also while we were at Waikura once, the water pump played up, so Hughy Hughes the Ruatoria Electrician flew in fixed it and then flew back. It was cheaper because of the distance to travel by road.

On the camp jobs we would often employ Willie the Cook from Tikitiki. The boys loved his cooking. He would do bacon and eggs say for breakfast but would cook it in a layer of fat. It used to turn me right off and I would just have a coffee or some toast.

At one of the Waikura camps I thought I would be smart and buy enough stores and booze to last a few weeks. But Muss could not stand the idea of having a couple of beers a day and invited all the station hands over for a party. I said fine but when it is gone that is it for a few weeks.

At another Waikura camp we had the mobile generator for power which was good because I was the boss and when I wanted to go to bed it was just a matter of turning off the generator and that was it, pitch black. No street lights out in the bush just a few wild possums running around.

In the summer the night skies were fantastic as you had no light pollution and just clear skies. The Southern Cross and the Milky Way look fantastic in the big open sky.

The life and times of Bill Rosoman and his involvement in the IT Industry.

2012 Update

Bill worked for the Telecom Corporation of New Zealand (previously the Post Office or Post & Telegraph) for 23 years until 1988. His time at Telecom was as a Foreman/Senior Foreman in the Construction and

Maintenance Branch involved in many facets of the Telecommunications Industry installing and maintaining the Outside Plant.

In the 1960-75 era the IT industry consisted of wind up telephones, wires and poles. It was mainly manual work digging holes and putting in poles and wire, using horses in difficult situations. There was two types of telephones available they were both black, there was a wall model and a desk model. People in Wellington and bigger centers waited many months sometimes years to get a telephone connected.

There were computers as such but they were very large IBM mainframes which took up several floors of office space, they were only available to large companies and governments. I used to work on these big mothers supplying them with links to the outside world.

In 1983 Bill brought a Sinclair ZX81 computer which plugged into the back of a TV for a screen and into a tape recorder for storing data on cassette tapes, the computer had a memory of 16kb which was upgraded to 64kb. The ZX81 used BASIC as its language.

Bill then brought a Sinclair Spectrum which was the second mass computer ever made. Next was a laptop of the XT variety, then a 286 upgraded to a 386. He then had a 486 notepad and then a Pentium 75 upgraded to 40mb of RAM and a 3gb Hard Drive. In 2000 Bill brought a 266mhz Notepad, and also uses a dual floppy XT Laptop running on Solar Power (twelve volts). In 2004 Bill has several computers. A Desktop 2ghz Intel (Mandrake Linux), a ASUS Notepad 2.8ghz Intel

(Windows XP + Ubuntu Linux), several old computers and notepads.

Bill purchased his first modem for the Spectrum, from Exeter in England as they were very expensive in New Zealand. It was a 300bps modem and the only person I could communicate with was Herby Kaa the then dentist at Te Puia Springs, some 10km, as he was the only other person I new had a computer and modem. The first major online event was called Videotex were I could view the weather, news headlines and my bank balance with Westpac. Then there was Telecom's Starnet and a Database in America that required the computer to dial 30 digits to get online. Now we mostly know about the Internet. Bill's first email address was rde002 and was on Starnet.

Bill has little formal IT qualifications and has learned computers mostly by sitting down and pressing buttons, reading books, asking questions and general perseverance. However in 2003 Bill Obtained a NZQA National Diploma in Computing stream Support Level 5.

NATIONAL DIPLOMA

Computing

Level 5

This is to certify that in December 2003

WILLIAM JOHN ROSOMAN

was awarded this qualification

Awarding Provider:
Advanced Training Academy Limited

Karen Van Rooyen
Chief Executive
New Zealand Qualifications Authority
Issued: 31 March 2004

Bill has been involved in the running of Internet sites at www.creativekiwis.com www.webng.com/leftfieldnz and www.webng.com/writernz These sites are housed in the USA. You might say that is his office.

He is also involved in Maintaining Websites, Writing Books, Software, and his specialty is telecommunications and in particular the Internet.

Bill was self-employed as a Computer Consultant from 1988 to 2002. Mainly in the Gisborne/East Coast region of New Zealand. Currently Bill now lives in the Waikato.

Bill believes that IT is one of the few industries where age, sex or knowledge and skills are not a barrier. Technology and its uses are only limited by the imagination and willingness to learn of the user.

Asus Z53J Notepad 2008

Bill has now had many years using Linux OS in particular Ubuntu Linux.

Ubuntu is all free and runs well. It does not get viruses or spyware and all the programs are free.

Bill Currently uses WIFI (Mostly at libraries), a 2 Degrees 3G Modem Stick and his Cellphone for Telecommunications.

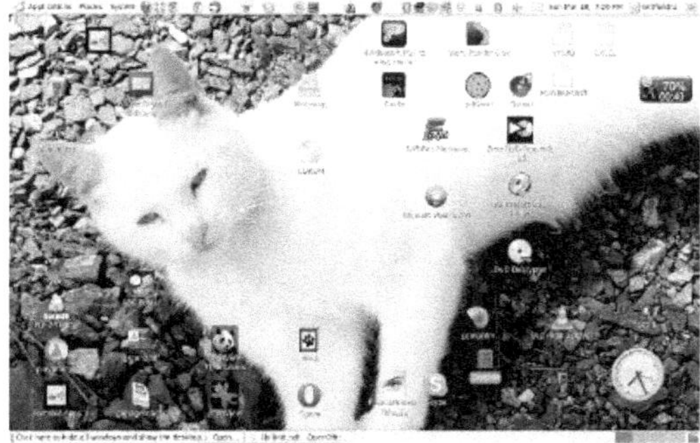

Desktop Ubuntu 10.04 2012

My current computer is an Acer Aspire 5920G, 4gb RAM, 250gb Hard drive.

I now also have an Android Minibook, which is great fun.

If you have any questions or would like further information please contact:

Bill Rosoman Dip CS leftfieldnz@gmail.com or Phone 021-233-5427

Other websites;

Personal Site

www.webng.com/leftfieldnz

Photos:

http://picasaweb.google.com/leftfieldnz

Movies:

www.youtube.com/leftfieldnz

Business Site

www.creativekiwis.com

The Evolution of the Internet

I started my career at Wanganui NZ, my home town, in 1965 as a Junior Lineman for the New Zealand Post Office. It was formally known as the P&T or the Post and Telegraph Department or the Puffed and Tired. I spent time also in Wellington on temporary transfer making lots of money but paying sixty cents in the dollar tax. It was a time were you waited quite happily for three months or more to get the telephone connected and then you had a choice of a black phone, desk type or wall mounted or nothing.

In 1970 while on my big OE in Europe, I was promoted to Tokomaru Bay as a Foreman Outside Plant.

In 1974 Tokomaru Bay the Telephone Exchange went Automatic and I moved to Ruatoria were I was promoted to a Senior Foreman and stayed till being made redundant in 1988.

When I arrived in Tokomaru Bay the phones were all manual wind-up jobs. I was the Foreman Faultman or in charge of the two man Telephone installing gang. It was quite a shock to the system!

We were responsible for all the Telephone outside plant of underground cables, telephone poles and wires and radio masts and aerials.

At Tokomaru Bay there was a Postmaster, a Post Office/Post Bank staff of about four people, there was the Telephone Exchange manned 24 hours a day by about ten staff, some part time and the Line staff of a Senior Foreman and Line gang of four or five, an Installing gang of two and one Faultman. The Area in Tokomaru Bay was from south of Tolaga Bay, almost to Whangara and to the North of Waipiro Bay.

That is an area of some 4,000 square kilometres. As the Faultman we had a four wheel drive Landrover but no maps, no radio, no cell phone, no GPS, just a lot of local knowledge.

The telephone exchange behind the Post Office was a manual exchange manned 24 hours a day to answer calls and put them through. Every call had to be put through on a cord, one end plugged into person making the call the other end plugged into the number to call and a switch flicked to ring the phone. And of cause there were the notorious party lines with four to six households on one line. Some of the poles and wires went for thirty kilometres or more to places like Huiarua Station on the Mata Road. It could be a nightmare when you had a fault on the line and a lot of walking the hills to find the problem. The only real aids were local knowledge, a pair of binoculars, some basic testing equipment and a sturdy pair of boots.

If I wanted to call my parents in Wanganui, I had to wind the handle and get the Operator at Tokomaru Bay, they would put you through to Ruatoria who would manually ring Napier who would ring Palmerston North who could dial my parents on an automatic line to Wanganui. Quite a mission as most of the way it was just a pair of copper wires (200lb/mile) on two cross arms on poles and that is if all the operators were talking to each other and not having a tiff, or were not busy. There was only a handful of circuits from Tokomaru Bay to Ruatoria.

In 1974 when the telephone Exchange at Tokomaru Bay went Automatic we were transferred to Ruatoria as they closed the depots at Tokomaru Bay and Tikitiki and built a new one in Waiomatatini Road, Ruatoria. There was a staff of up around a dozen. One Overseer, One Senior Foreman and four man line gang, two Foreman faultmen, one Foreman and Lineman installing gang, one storeman/cleaner. There was also a large staff at the Post Office and Manual Telephone Exchange.

The Ruatoria Telephone Exchange went Automatic in 1977, I remember connecting the Lotto at the local four square grocery store as it was the last Lotto to be connected in New Zealand as I understood.

Ruatoria Phone Book 1950s or 60s (this was the complete phone book).

OFFICIAL TELEPHONE DIRECTORY. 37

NUHAKA—Continued.

Exchanges and Bureau Stations with which communication can be obtained
(For Charges see Inside Front Cover.)

Bureau or Exchange	Code Call	Miles Distant	Hours of Attendance	Half-holiday Day	Hour
Morere	MRR	5	9 a.m. to noon and 1 to 5 p.m.	Wednesday	1 p.m.
Opoutama	OPM	7	9 a.m. to 5 p.m.	Wednesday	1 to 5
Tongitera	TRC	15	9 a.m. to 5 p.m.		
Wairoa	WA	26	7 a.m. to 11 p.m.		
Wairoa Bureau Stns			8 a.m. to 5 p.m.		
Whakaki	WKZ	11	9 a.m. to 5 p.m.	Saturday	12

RUATOREA EXCHANGE (85 miles).
Hours of Attendance: 9 a.m. to 5 p.m.

15M Awatere, Pehikura, 'Nagahanoa'

20M Beach, H., 'Ruatorea'
20S Beach, R., 'Waimoka'

14M Grace, Wm. R., 'Tuparoa'
53S Grace, Samuel, 'Kaimoho'

14D Haenga Kerau, Waitekaha
14R Haenga Ruka, Tuparoa
7M Holt, S. J.
7W Hospital, 'Manutahi'
37 Hotel, Manutahi
54R Hotel, Tuparoa

31M Jefford, C. T., Omaramuai
17 Johnston, W. H. Owen, General Store, Ruatorea

33M Kaiwai Parepare, Tuparoa
33S Kaiwai Tamati, Taumataomihi
38S Kaiwai Timoti, Takatotore
23R Katia Rawiri, Taumataomihi
31S Kemp, F. J.
36M Kirk, A. W., Tuparoa
34D Kirk Bros., Tuparoa
34S Kirk Bros., Ruatorea

32D Ludbrook Bros., Waiorongomai
42S Ludbrook, E. R., Tuparoa
12S Ludbrook, E. R., Tuparoa
42M Ludbrook, W. W., Waiorongomai

13M McClutchie Ken, Mahora
57M McClutchie, W., Mangawhariki
59 Maki Paku, Hiruharama
25 Manutahi Hall
43D Macke Erueta, Whakahu

63R Ngata, A. T., Kakariki
98M Ngata, A. T. Hon., Waiomatatini
90S Ngata Paratene, Waiomatatini

58R Pakaroa Hall
55D Pakatai Ehau
14R Paku, Hall, Tuparoa
56R Pokai Hakare

14W Rairi Hohepa, Tuparoa
80M Reed, A. V. S.

Sub-Exchanges: Nuhaka, Ruatorea, Te Araroa

38 OFFICIAL TELEPHONE DIRECTORY

RUATOREA—Continued.

80R Reed, A. V. S., Pakihiroa
57W Reedy, John W., Koputa
83W Reedy, John W., Wharekōpae
23M Rewarewa Hekiera, Kariaka

24 Smyth, H. F., Storekeeper, Ruatorea
13S Sullivan, M., Mahora

14S Tamaheri Pine, Tuparoa
15R Tamaumbi, Piniha, Ruatorea
4 Taumaunu Matahara, Ruatorea
18S Tawaho Wi, Mangakanea
58M Te Rapu Hami, Te Raupo
12W Tuparoa Station Office, Tuparoa
12D Tuparoa Station Store, Tuparoa

40 Wakarara, D.
56D Watkins, B. C., Ohinepoutea
56R Watkins, Kenneth
86S Williams, C. T., Mohoiwi
65R Williams, K. S., Matahina
59W Williams, O. T., Reparapoririki
32 Williams, T. S., Kaharau

Exchanges and Bureau Stations with which communication can be obtained.

Bureau or Exchange	Code Call	Miles Distant	Hours of Attendance	Half-holiday Day	Hour
Kahukura	KAU	12	9 a.m. to 5 p.m.		
Te Araroa	TRL	28	8 a.m. to 8 p.m.		
Te Puia Springs	TPS	18	9 a.m. to 5 p.m.		
Tokomaru Bay	TKY	25	7 a.m. to 11 p.m.		
Tuparoa	TOA	9	9 a.m. to 5 p.m.		
Waipiro Bay	WIP	15	9 a.m. to 5 p.m.		

TE ARAROA EXCHANGE (103 miles).
Hours: 8 a.m. to 8 p.m. Closed on Sundays and Holidays.

60 Abraham, L. H.
18W Akapu, Te Paru
10K Akuhata, Wiremu, Te Rimu
11K Allen, F. G., Pipituangi

21 Bank of New Zealand
9S Beale, J. H. (res.)
2K Beckett, C. L. B.
2M Black, William (res.)
28 Bousfield, F. (res.)
15R Bristowe, Henry
15M Bristowe, Heneta
24R Brown, Mrs. George (res.)

18K Clarke, C. H.
3M Clarke, J. H.
14R Cunningham, J. A. (res.)

24D Fagan, W. S. (res.)

20M Goldsmith, C. (res.)
11W Goldsmith, George
18S Gudgeon, Herman & Melville

13R Halliwell, M. W., Ewdrow
19K Hamilton, H. A. A.
70 Hansen, C. L.

An Early Email (1989)

```
R ALL

      To:  RDE002
    From:  ROSS.K  (TLC424)   Delivered:  Thu  23-Nov-89  17:06 NZST SysY 6
401  (11)
Subject:  PACKNET ACCESS
Mail Id:  IPM-6401-891123-153920686

PACNET MISSPELT NEVER MIND

FROM STUART JOHNSON P BUS SALES
ACCESS TO RUTORIA SHOULD BE AVALIABLE AND HAS BEEN
FOR A FEW WEEKS NOW.
EXCHANG STAFF HAVE DONE A PURGE.
PLEASE ADVISE IF ANY FURTHER DIFFICULTIES ARRISE WITH
ACCESS
.



      To:  RDE002
    From:  STARNET.FAX  (NZFAX) Delivered:  Mon  27-Nov-89  14:05 NZST Sys 6401
(9)
Subject:  STARNET TO FACSIMILE
Mail Id:  IPM-6401-891127-126760566

CONFIRMATION OF FACSIMILE DELIVERY

Destination   07945799
Destination area   Domestic
Transmitted   89/11/27 14:06:18     Mail Id: IPM-6401-891127-125950172

Number of Pages   1
Total Price (Excl GST)   $1.80


    From:  VIVIAN.H  (CAT001) Delivered:  Wed  29-Nov-89  7:11 NZST Sys 6401
(124)
Subject:  FROM VIVIAN HUTCHINSON
Mail Id:  IPM-6401-891129-064650103

        HELLO ALL FRIENDS
        AS THE FOLLOWING NOTE EXPLAINS, WE ARE CLOSING DOWN THE PUBLIC WORK OF
    THE RAINBOW CATALYST BY THE END OF THIS YEAR. THIS MEANS ALSO THE CLOSURE
    OF OUR "CAT" USER GROUP ON STARNET.

        IAN RITCHIE AND I HAVE JOINED UP WITH THE "ERN" GROUP OF STARNET
    USERS, UNDER OUR RESPECTIVE RESOURCE CENTRES. SO IN THE FUTURE WE ARE
    CONTACTABLE THROUGH :

        ERN026 IAN RITCHIE and ANGELA BAKER
               MANAWATU RESOURCE CENTRE, PALMERSTON NORTH

        ERN027 VIVIAN HUTCHINSON and DAVE OWENS
```

In the 1960-70 era the IT industry consisted of mostly wind up telephones, and wires and poles. It was mainly manual work digging holes and putting in poles and wire, using horses in difficult situations.

If you were rich or important you would have a Telex Machine (these died in the 80s) and later a Fax Machine. The biggest use of Telecommunications apart from the Telephone was sending and receiving Telegrams, especially as they were free to Post Office staff.

There was some Radio Communications via CB (Citizen Band), Amateur Radio, Shipping and Trucks etc. and Shortwave Radio.

I listened to shortwave radio from the 1960s to the 1990s

This is a QSL Card from the Voice of America.

My Barlow-Wadley Shortwave Receiver.

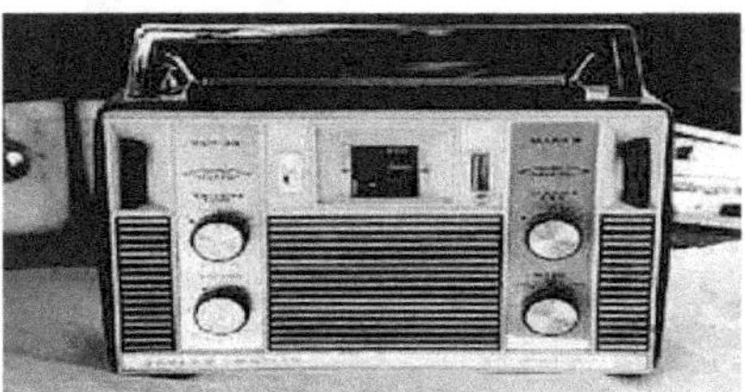

There were computers as such but they were very large IBM mainframes which took up several floors of office space, they were only available to large companies and governments. I used to work on these big mothers in Wellington supplying them with links to the outside world. IBM famously said there was only room for half a dozen computers in the world.

The Post Office Computer at Herd Street Wellington as I understand just did the pay for 35,000 Post Office staff, took up a whole floor and they dug a big hole in the yard to bury a water tank and diesel tank for standby power etc. It was huge and limited in it's application.

In 1983 I brought a Sinclair ZX81 computer which plugged into the back of a B&W TV for a screen and into a tape recorder for storing data on cassette tapes, the computer had a memory of 16kb which was upgraded to 64kb. The ZX81 used BASIC as its language.

Sinclair SX81

I then brought a Sinclair Spectrum which was the second mass computer ever made. Next was a laptop of the XT variety, then a 286 upgraded to a 386. He then had a 486 notepad and then a Pentium 75mhz upgraded to 40mb of RAM and a 3gb Hard Drive. In 2000 I brought a 266mhz Notepad, and also used a dual floppy XT Laptop running on Solar Power (twelve volts).

I purchased my first modem for the Spectrum, from Exeter in England as they were very expensive in New Zealand. It was a 300bps modem and the only person I could communicate with was Herby Kaa the then dentist at Te Puia Springs, some 10km, as he was the only other person I new had a computer and modem.

The first way to get on-line was with Bulletin Boards. These consisted of Computer Enthusiasts setting up a Bulletin Board for others to access via a modem. There was a guy at the Bank of New South Wales (Westpac) Bank who set-up a Bulletin Board in Gisborne. I used to dial into it by making a toll call and loging on. You could access some software, forums and could send and receive email. This Bulletin Board would contact another in say Auckland or Sydney say once in every 24 hours and exchange information. This would then bounce around till reaching its destination. So an email and reply could take several days to get there and back to you. Of cause all this was in DOS with primitive programs. I remember one of the first email programs was called SLMR (Silly Little Mail Reader). We also used Software called Prism to communicate with Bulletin Boards. You had to a bit of programming to get things to happen. It was not very user friendly. You could transfer files using protocols like Kermit, easy for geek types.

The second major on-line event was called Videotex were I could view the weather, news headlines and my bank balance with Westpac. You could only look at stuff on-line, you could not reply and the screen of cause was Green or Amber. Videotex was run I think by the Government Computer people. I forget what they were called.

Then there was Telecom's Starnet and a Database in America (McDonald Douglas Super Computer) that required the computer to dial 30 digits to get on-line Now we mostly know about the Internet. My first email address was rde002 and was on Starnet. Starnet and Pacnet could cost round $200 per month.

To access Starnet you had to use Pacnet. So you required an account for each. Pacnet was an access gateway to many databases and computer systems like Starnet.

The 1990s I had Intel Pentium computers, with colour and multimedia capabilities so necessary in today's IT world. Windows 3.1 (which came on about 18 3.5" Floppy Disks) We had a 14kbs then 28kbs, then 33kbs and finally 56kbs modems and Dial up Internet connection and e-mail addresses. We could finally multi-task.

The Internet as we know it in 2009 came in the 1990s. First I used to access CompuServe via a toll call and then log onto CompuServe at many cents per minute. I once had a monthly bill of $500 and was usually in the hundreds of dollars. I made a site at Geocities www.geocities.com/leftfieldnz in 1995 and acquired a Yahoo email at the same time leftfieldnz@yahoo.co.uk

I next managed to get an email address through the Tairawhiti Polytechnic who had a pair of wires and a 64kb link to Waikato University who were the gateway for New Zealand Internet at that time.

Finally we got access to the Internet proper via dial up on a 9.6kbs Modem and an 0800 number.

At first people could not get there heads around having modern 24 hour a day communications. In Tokomaru Bay there was classic examples of the misunderstanding of technology. From the Post Office Union saying we had to fight and stop technology if we wanted to keep our jobs. I said to the Union guy, that if you have flown all the way from Wellington to tell us that you are wasting your time, technology is going to happen. They made Telecom make the Telephone Jacks sealed, so you had to get a technician in move, replace or install a telephone.

The locals also called a public meeting to stop Westpac, the last bank in Tokomaru Bay from closing. I was the only one saying close the bank as we have technology like the Internet and EFTPOS cards and they were all saying no we need to keep the bank open. It was only open half a day a week anyway so a waste of time. Westpac closed and the locals used EFTPOS.

I went to the Potaka School in remote East Cape and gave them a demo of the Internet, I asked a guy his car registration number and had a look on-line and could tell him he owed money on the car, he said how do you know that. I said you could now read the paper on-line when you want instead of waiting for the bus in the late afternoon to deliver the paper.

On another occasion I went to Pewhairangi's as Cody wanted a printer driver. I put my laptop down in the middle of the lounge and plugged in my laptop to the power and the phone line, she said, what are you doing and I explained I am using the Internet to get a driver for the printer. I installed the driver and She was impressed! Before this you had to ring the manufacturer or importer and plead with them to send you the driver on a floppy disk.

I was working at Nga Roimata Training Centre and Wiki used to pay the wages by cheque, you the had to bank the cheque and wait five working days for the money to be available, or as Wiki used to do, drive 40km to Tolaga Bay cash the cheque and put into your account. I fought to get ASB Desktop Banking and it was brilliant. I would work for the day, give Wiki an invoice, they would pay me on-line and then I could walk straight to the shop and access the money via EFTPOS.

Also at Nga Roimata, we wanted to order some educational material from a catalogue we had. I made the order out then rang the people concerned in Auckland to find out there bank details so I could make a direct credit to their account and then fax the order through to them. Well the sales person could not handle that. I was put through to the accountant who said their bank details were confidential. I said can you shop 24 hours a day on-line and do all the things the Internet has to offer, it took him awhile to figure out what I was saying but in the end gave me the account details. He said but how will we know it is from you I said we will have our name on it and it will be for $256.11 or what ever the amount was. He said this will be a first, I said welcome to the 20th Century.

My late friend Kevin Makin was a researcher and suffered from epilepsy. If he wanted information from the National Library. He had to hitch or take a bus to Gisborne, fly to Wellington, access the Information and then return, taking 2-3 days and costing hundreds of dollars. Now we could do research and contact people as well as do things like on-line banking from Tokomaru Bay. His Company was Research Design hence my email rde002 on Starnet. We lived about 1km apart. One Sunday I was doing a document and wanted Kevin to have a look at it. Normally I would have put it onto a Floppy Disk taken down and got Kevin to have a look. But I rang him and said look the weather is nasty will you check your email and I will send the document to you for you to look at, edit and return to me. Marvellous, but how we take it for granted in the 21st Century.

I have little formal IT qualifications and have learnt computers mostly by sitting down and pressing buttons, reading books, asking questions and general perseverance.

However in 2003 I obtained a NZQA National Diploma in Computing stream Support Level 5.

I have been involved in the running of Internet sites at www.webng.com/leftfieldnz and www.webng.com/writernz These sites are housed in the USA. You might say that is my office.

The latest project in 2009 was www.creativekiwis.com a website to market some of my Books and DVDs etc.

I was also involved in writing software, computer training and my speciality is telecommunications and in particular the Internet.

I was self-employed as a Computer Consultant from 1988 to 2002. Mainly in the Gisborne/East Coast region of New Zealand. Currently I live at Hamilton in the Waikato.

I believes that IT is one of the few industries where age, sex or knowledge and skills are not a barrier. Technology and its uses are only limited by the imagination and willingness to learn of the user.

I now have had several years using Linux OS in particular Ubuntu Linux. It is all free and runs well. It does not get viruses or spyware and all the programs are free.

I also brought an old laptop and put Puppy Linux on it, it is fantastic. Puppy Linux basic is only 126mb and will run on any old computer. Either running from a CD or installed on a hard drive, you have all the basic to get up and at them, even the internet and all. Sad the laptop died!

I have an American web-pal who tells me we have been corresponding for 13 years. That has been very nice.

A not so typical set-up in 2009 is at Glen Massey in the Waikato. They have Satellite TV (Free to Air) and Satellite Internet as well as Solar, Wind and Diesel for Power. The Internet cost $2700 to install and $69 per month for a 2gb data allowance.

Mind you it can be a pain, as the Internet is only available if there is power. They use cell phones for immediate communication and Internet VOIP like Skype for long distance calls.

BTW, Took this Photo with my Cell Phone and transferred the Photo to my notepad Via Bluetooth.

You know wonder in the 21st Century how we managed before Broadband Internet or Dial up Internet.

All this has happened in my brief 60 years on the planet!

Getting Broadband in NZ in the 21st Century,

aka working with Telecom NZ, Telstraclear and Slingshot.

or Life in the Dark Ages under the "Teleban"!

What a freaking saga!!!!!!!!!!!

My questions,

In the 21st century with all our "sophistication and technology" how come there is not a seamless process for dealing with telecommunications companies to get good quality and on time service, like broadband?

How come all the telecos seem to go out of there way to make life difficult and do not see the travesties and frustration that they cause, and do not get it (they are the problem)?

How come the telecos are unable or refuse to cooperate with each other for the benefit of the fee paying customer. the person they all make mega bucks out of?

How come the telecos can not see there is a problem and fix it seamlessly?

How come slingshot can detect I am no longer using them as my primary toll provider (because we have been cut off), and penalise me accordingly, but cannot do the same with our broadband connection and fix the bloody thing?

How come in the 21st century we are incapable of getting it together and acting like 21st century people and not like we were still in the dark ages under the "teleban"?

Do these Teleco companies practice insulting customers as an art form, talk about adding salt into the wound!!!!!!!!!

The Saga

NZ Telecom in NZ for 20 years has had a monopoly on telecommunications and has had a wonderful time making exorbitant profits of around $900m a year, paying exorbitant salaries, like a CEO on $1.7m plus all the perks (let alone the wonderful golden parachutes they all seem to get), and doing all it can to frustrate competitors and regulators.

It all goes back to Rogernomics in NZ in the 1980s were Telecom was "sold" three times. First it was the NZ Post Office a government department were you waited many months to get a phone on, you had the choice of a black phone or nothing and were happy about it. Unless of cause you got a P1 status, i.e. a Post Office employee, a diplomat or had a serious illness etc., when you got the phone on in 24 hours. Any way NZPO was made into and SOE (State Owned Enterprise) and sold for around $4b. This "sale" was of something that the government did not even really own as it was owned by the "subscribers". A couple of years down the track labour decided that was not good enough and sold it to American private interests for around $7b. The Americans then floated it on the stock market and eventually sold up and left town!

So something the government never owned was flogged off three times, but the biggest "crime", was labour MP one Richard Prebble, selling

Telecom and the Numbering system. We are the only country in the world to sell off the numbering system to private interests. This means that 20 years on we have no number portability or unbundling of the loop

Number portability is the ability to take your number were ever you go and using it with any telecommunications company. Unbundling, is not having to have the telephone and broadband via copper wire with the same company, or having to have a telephone at all.

In 2006 the labour government finally got some some balls and decided to deal to Telecom NZ and split into three operations, to get some competitiveness into the whole telecommunications industry in NZ.

So that leads to 2007 and my attempts to get broadband internet at our home in the Waikato.

In November 2006 I spent some $1300 buying new technology and made round $600 selling off some of our older technology. This included a wireless router with a built in ADSL modem. I installed these and put Linux Kubuntu on all the computers (one desktop and two laptops). This works great with Wifi as you get broadband and a WLAN network all at the same time as opposed to Windows XP Home were it is almost impossible to network the computers (only available in XP Pro). By November 2006, things were finally beginning to loosen up to the point were I could get broadband with no 12/24 month contract. So after studying all the deals I decided www.Slingshot.co.nz had the best deal.

Currently I was paying $16.95 per month for dial up internet, $40 per month for a telephone line and tolls (long distance) on top of that. Slingshot broadband was $39 per month plus the phone through Telecom of $40/month. I justified at it as slingshot had a toll deal of $2 for 2 hours calls. I figured all added up I was paying much the same as it was.

BTW if I was in Canada or the UK I could get 30gb 2mps broadband for $30/month.

Now I should also say we had the actual telephone connected through Telstraclear. So in November I spent most of one day trying not too lose my cool and talk to Slingshot, Telecom and Telstraclear to get broadband connected. We had to change the telephone back to Telecom as slingshot "did not have an arrangement with Telstraclear". Slingshot could then not talk to us because we did not have a Telecom account number "for security reasons". So we had to talk several times to Telecom to get the connection process started and an account number so slingshot would start their process of connecting broadband.

The other funny thing was that I had done some research and found that ADSL broadband will only work on a few kilometres on copper wire, I asked the Telecom guy were the local Whatawhata telephone exchange was and if we were in range. He said he could not tell me "for security reasons" despite the fact I could probably find it in few minutes on Google Earth.

Still in early December 2006 we finally got broadband all be it at 256kb download and 128kb upload and a cap of 25gb per month (hardly world breaking speeds and a joke in the 21st century).

There was some hiccups getting connected at times and slingshot said "it was a known issue with Telecom, and it was being worked on".

Sigh, so by this time the average little old lady would be suicidal or being committed to the loony bin!!!!!!!!!!!

After one month we got a Telstraclear phone bill as normal, except we not with them any more. I rung Telstraclear but they refused to talk to me as I am a mere male. IE the telephone is my Nieces name and if I had been female and given her name, address and date of birth it would have been accepted for "security reasons". So my niece rang and was told there was no one there of the holidays to push the button and we have to wait.

On Wednesday 4-1-07 my niece rang telstraclear to cancel the telephone we no longer had, which lead to the telephone being completely disconnected that day. Sigh, my niece rang Telecom on the 5-1-07 and after an hour got the telephone reconnected, unfortunately she not knowing all the technical issues and the incompetence of Telecom and Telstraclear, we wound up with just the phone and no broadband or tolls with slingshot.

On Friday 5-1-07 I rang slingshot and was informed there was nothing they could do till the 8-1-07 "as with normal business practice, all our staff are on holiday", to which I said you are joking, you do not have one person on who can push the right button! But alas she was not joking, but did give us our dial up back till they sort it out.

On Saturday 6-1-07 I got an email from slingshot saying "we note that you no longer are using slingshot as your primary toll call provider, and we are going to charge you more next month". Shoot by now you gotta laugh, or you would blow a fuse. So I rang slingshot at 8am only to be told after being disconnected twice that technical staff do not start till 10am on the weekends.

Yippee the Saturday mail has just arrived which includes a ADSL modem from Telecom as they are trying to get us to do a deal with them!

Rang slingshot at 10am Saturday 60107 “2nd in cue”, I was then asked to wait and was put through to the technical support people and a recorded message said “thank-you for calling technical support, our hours are 8am to 10pm Monday to Friday and 10am to? Saturday and Sunday” and then I was disconnected.

Rang back and was disconnected, rang back again, for the first time I am told I am ringing new sales and should ring a different number 0800-89-2000 and select option 3 then 1 then 3. It is 3, 2, 2. Great music! 15 minutes later, I am 4th in cue,. After 40 minutes I get through, they have to have an account number, my name, telephone number, logon id, will not do. We get to discuss my problem, after my calls yesterday the connection request has gone out to the team that connects broadband at slingshot, but I have not been scheduled for connection, I can not find out when that might happen, but it might be today (Saturday). I will not hold my breath!

The saga continues??????????????????????????????

This rave is in total frustration and desperation and so I do not become suicidal!!!!!!!!!!!! LOL

Bill Rosoman

Communications Online Telecommunications Online

Using Skype Yahoo Messenger etc.

Yahoo Messenger..... Skype VOIP

Yahoo Messenger Conference Call .... Skype Conference Call

You and your Friends/Associates only need to have a Computer with access to the Internet

(Dial up is OK, but Broadband is better), or access to a computer or if the worst is to be a Phone or Cell Phone.

And then you can Chat, Share Webcams, Photos, Files Etc.

Yahoo Messenger

One way to Communicate and have Conference Calls Online, is to use Yahoo Messenger.

You can download the required software for Windows/Linux/Mac and use it that way (Picture on Left). Or you use it from a webpage @ meebo.com (Picture on Right). Or from a 3G mobile

Click on an image to goto Yahoo Messenger or Mebo,

You have to have a Yahoo ID and Password (you can sign up on their website and it is all free). My ID is leftfieldnz. Then you login and can chat to friends and do dozens of other things like share conversations, webcams, photos, files etc. etc.

Or you can have a Conference Call.

You can even use Yahoo Messenger on a mobile

Even when you're out, you're in.

Yahoo! Messenger keeps you connected even when you're away from your PC. There are two ways to use Yahoo! Messenger with your mobile phone:

Send IMs as text messages from your PC to your friends' mobile phones

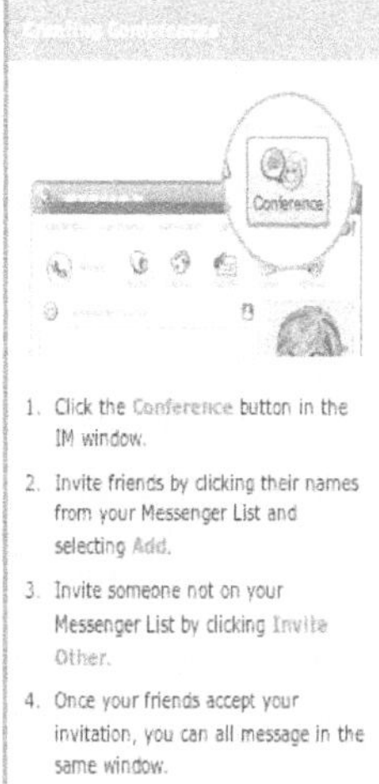

Sign into Yahoo! Messenger from your mobile phone and send IMs to your friends when you're away from your PC

To start using Yahoo! Messenger right from your phone, find out which version works best for your phone model, it's easy!

Go to Yahoo! Mobile

Choose your carrier Enter your phone brand and model

Get Yahoo! Messenger on your mobile phone and get going!

Yahoo Conference Call

It's easy to have a conversation with multiple people when you use Conference in Yahoo! Messenger. Communicate via text in one window, or share your webcam with several people at once.

(1) Enable video by clicking the Webcam icon.

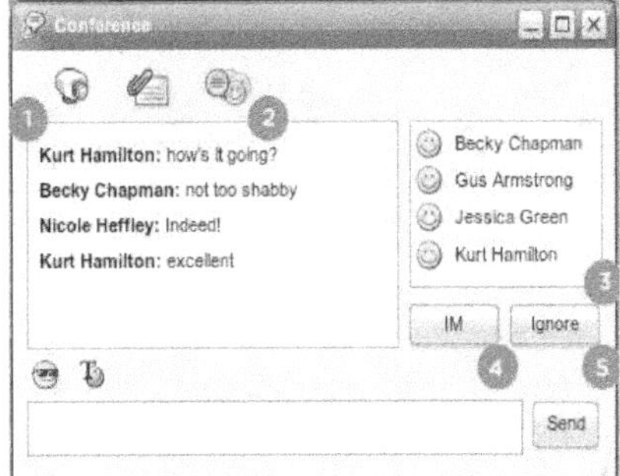

(2) Add more friends to your Conference by clicking the Conference icon.

(3) The conference participants are listed to the right of the conversation.

(4) Send a friend a private IM by highlighting their name and clicking IM.

Ignore a friend's messages by highlighting their name and clicking

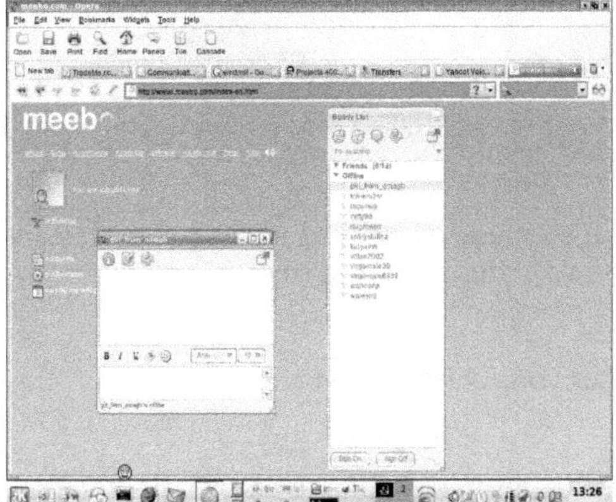

Ignore.

Skype VOIP

www.skype.com

Another way to Communicate and have Conference Calls Online, is to use Skype VOIP.

You can download the required software for Windows/Linux/Mac and use it that way (Picture on Right).

You can use Skype for calls Online and to Landlines and Cellphones. Computer to Computer is free.

You can call people who only have a landline or mobile. Although this costs (You need to Buy a 10 Euro card from Skype and it cost around 3 cents per minute to NZ and more for mobiles).

You have to have a Skype ID and Password (you can sign up on their website and it is all free). My ID is leftfieldnz. Then you login and can chat to friends and do dozens of other things like share conversations, webcams, photos, files etc. etc.

Skype Conference Call

Open Skype and Logon if asked.

Goto Tool Create a Conference Call. Add the Friends you wish to Conference with and press Start.

I am pretty sure you can include people who only have a landline or mobile. Although this costs (You need to Buy a 10 Euro card from Skype and it cost around 3 cents per minute)

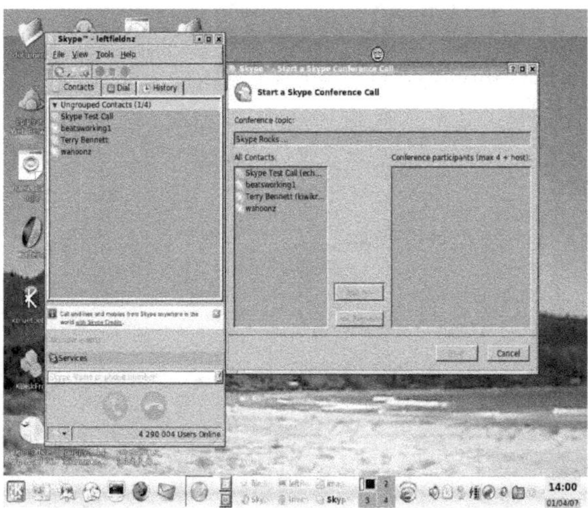

Mobile Communications on holiday or on the road

What You can do Online | Cheap Chat Phone Card | Mini Opera Web Browser | TXT Banking

What You can do With a Mobile | 2Degrees | Telecom | Vodafone | 2Degrees Setup

Mobile Communications on holiday or on the road

When on Holiday or not near a computer or the internet you can still keep in touch with some of these ideas.

Nowadays they call it Cloud Computing were you have a network device IE Cell Phone and can do all sorts of things with the one device, both on the Internet and just with TXT and Voice.

I suggest a reasonable phone (I had a Sciphone, an Iphone knockoff for $190 from www.trademe.co.nz), A 2degrees SIM Card and a Cheap Chat Phone card. Mine is a Dual SIM Phone so I have a Vodafone SIM as well. I now have an Android Smartphone.

2 Degrees is brilliant. When you top up with $20 voucher you get 30 days at super rates. A TXT 4c Voice Call 22c/min. Not for power uses but great for casual users.

I have been on holiday for a week and it has cost me $4.

Helpful Websites

m.google.com/?hl=en

m.gmail.com/

www.facebook.com

m.2degreesmobile.co.nz

www.vodafone.co.nz

wap.oa.yahoo.com/#c1

news.bbc.co.uk/1/low

www.kiwibank.co.nz/

www.radionz.co.nz/news

m.trademe.co.nz/

www.mysciphone.com/

www.yourwap.com/

maps.google.com/

docs.google.com

www.twitter.com

www.bebo.com

www.meebo.com

m.nzherald.co.nz

www.telecom.co.nz

www.youtube.com/leftfieldnz

picasaweb.google.com/leftfieldnz

www.webng.com/leftfieldnz/wap3.html

www.creativekiwis.com/

www.tinyurl.com/

www.travelite.org/

Some of the things you can do Online

Google

Email

Calendar

Google docs

Picasa Pictures

Youtube Videos

Maps

Search

Facebook

Upload/Download Photos/Videos

Chat

Online Office

Cheap Chat Phone Card

www.telstraclear.co.nz

A Cheap Chat Phone Card or similar is great as you can make calls from any Telephone to anywhere.

From a landline to a Landline 3c min

50c min to cell phone

0800-552-885

or

0508-508-773

Mini Opera Browser

Download the Mini Opera Browser to your Cell Phone, Install and Use as Internet Browser.

www.opera.com

Banking using TXT (SMS)

Kiwibank www.kiwibank.co.nz

Register Cell Phone on Website

TXT BAL to 5494

ASB Bank www.asb.co.nz

Register Cell Phone on Website

TXT BAL to 4272

How to register

Login to FastNet Classic

Select the 'Mobile Banking' menu

Select 'TXT banking' and follow the steps on screen

When you register, you choose up to three of your accounts that you want to access through TXT banking. Each of these accounts is then given a FastCode, which you use as the account name when TXT banking. You must be Netcode registered to complete registration on FastNet Classic.

You can change your selection at any time through FastNet Classic.

To get a quick balance

For all three accounts at once, TXT: Bal to 4ASB (4272)

For one account, using the account's FastCode, for example if this was 100, you'd simply TXT Bal 100 to 4ASB (4272)

To transfer money between two of the accounts

For example, to transfer $50.00 from account 100 to account 101, you'd TXT: Trf(space)$50(space)f100(space)t101 to 4ASB (4272)

The f represents the account the transfer is coming from, and the t being the account the transfer is going to.

Note: You cannot transfer True Reward dollars to another account or credit card nor make transfers to your True Reward dollar balance.

*This fee is in addition to any other ASB base or transaction fees with no exemptions, except FastSaver. Fees subject to change. Your normal mobile phone network provider TXT message fee also applies and will be charged to your mobile phone account or direct debited from your prepay account balance. Free TXT deals offered by your mobile phone network provider do not apply to ASB's TXT Banking service.

Things you can do with a Mobile Phone

Internet

Email

Web Browser

Voice Calls

Text Message (SMS)

Video Message (MMS)

Camera

Pictures

Video Recorder

Audio Books

E-Books

Video Player
FM Radio
Sound Recorder
Calculator
Currency Calculator
Address Book
Contacts
Games
Photo Album
Play Music (MP3)
Banking
Blue Tooth
Alarm Clock
Chat

2degrees Phone Numbers
0800022022
*100# Balance and Topup
*#06# Phone Serial Number
topup 201
help 200
voicemail 202

Telecom Phone Numbers
0800-000-000 or 0800-800-123 or *123
0800-32 32 32 Balance and Topup
*#06# Phone Serial Number
*333 For Topup

Vodafone Setup
0800800021

COUNTRY
New Zealand

CARRIER
Vodafone

Internet Settings
APN
live.vodafone.com
or
www.vodafone.net.nz

USER

PASSWORD

HOMEPAGE
http://live.vodafone.com

2degrees Setup

services
WAP
settings
edit profile
sim2
select profile
edit profile
Homepage
http://m.2degreesmobile.co.nz

Telecom APN

wap.telecom.co.nz

To Connect a 3G Modem Stick

Plug in Modem

When it appears on Desktop as a CD

In Windows just install the software on the 3G stick and connect.

In Linux;

Open Console

type dmesg

Should see ttyUSB0 etc.

type sudo wvdial

Enter Log On Details

Make sure Stupid Mode and Auto-Reconnect are Ticked

Make sure the Selected Modem is ttyUSB1 or 3 (the device is usually ttyUSB0, but the Modem part is ttyUSB1 or 3).

If not 1 or 3 Press Choose

Press Connect

My Modem is;
Phone *99#
APN internet
User 2degrees
Password internet

Smart Phones

What an Android Smart Phone Can Do.

Heaps is the answer.

You can do;
Ebooks
Audio Books

Music
Movies/Videos
Create Letter/Document
Create Spreadsheet
Compose a Presentation
Internet/Web Browsing
Email
WIFI
View Pictures
Skype Chat
Most of the Google Apps like Google maps
Games
Bluetooth
Camera
Calendar

Sync with Google
Alarm Clock
App Store
Camcorder
Contacts
Messaging
RSS Reader

Plus heaps of Apps like GPS etc. etc.

Installing Apps

The easiest way is to use the App Store by Clicking on the App Store Icon, Find the App You want and Download and Install.

Otherwise go to

https://play.google.com/store?hl=en
http://www.camangimarket.com
http://www.freewarelovers.com/android
http://www.google.com/mobile/android
www.slideme.org

Locate the App you want. Copy to SD Card and the install from that.

Converting Video and Audio for Android Smart Phone

I use several programs to convert media like movies into movies that will play on a smartphone or Tablet,

These packages are available in Windows and Linux.

Mobile Media Converter

http://www.miksoft.net/products.htm

From MIKSOFT:

The Mobile Media Converter is a free video and audio converter for converting between popular desktop media formats like MP3, Windows Media Audio (wma), Ogg Vorbis Audio (ogg), Wave Audio (wav), MPEG video, AVI, Windows Media Video (wmv), Flash Video (flv), QuickTime Video (mov) and commonly used mobile devices/phones formats like AMR audio (amr) and 3GP video. iPod/iPhone and PSP compatible MP4 video are supported.

WinFF

http://winff.org/html_new/downloads.html

WinFF is a GUI for the command line video converter, FFMPEG. It will convert most any video file that FFmpeg will convert. WinFF does multiple files in multiple formats at one time. You can for example convert mpeg's, flv's, and mov's, all into avi's all at once. WinFF is available for Windows 95, 98 , ME, NT, XP, VISTA, and Debian, Ubuntu, Redhat based GNU/Linux distributions.

With WinFF you need to install ffmpeg first.

ffmpeg (see guide later in book)

ffmpeg for windows

Download Windows ffmpeg binaries from http://ffmpeg.arrozcru.org/autobuilds/

In windows we use a back slash \. c:\bill In Linux we use a forward slash /. /media/Acer/

start (Menu)

run

cmd

OK

Type in cd \bill\ffmpeg

Type in

ffmpeg -i \bill\movies\100_2747.MOV -y -target pal-dvd -aspect 4:3 -b 1800 \bill\movies\mov1.mpg

Or the location of your movies

Use TAB to complete operation

like \bi then TAB will complete the rest \bill

Use the up arrow and change the video names and go again

cd ..

Takes you up one folder

http://www.zamzar.com/

Free Conversion site.

Converting MPEG to 3GP using FFMPEG (should work for avi etc. movies as well)

ffmpeg -i /media/ACER/1/2/1/movies/matilda.mpg -s qcif -vcodec h263 -acodec libfaac -ac 1 -ar 8000 -ab 12.2k -y /media/ACER/1/2/1/movies/matilda.3gp

Join Video and Audio

ffmpeg -i final.mpg -i sound.mp2 final_snd.mpg

avi to flv (some phones and tablets and Youtube)

```
ffmpeg -i movie.avi -s 320x240 -ar 44100 -r 12 movie.flv
```

All the above work for me!

I also use Calibre (see later in book) to convert to Ebooks.

Conclusion

When my Great Nephew was about 9 I Said, In my day boy, there was no TV, no Cell Phones, no Calculators in exams, no Computers, no internet, no email, no DVD players etc.

He looked at me and said, bull shit Uncle as he has no concept of not having these things.

How things have progressed in my lifetime.

Things are moving at a great pace, but mostly it is good.

I'm just enjoying the ride!

We would like to thank you very much for your interest in our books, ebooks and audio books.

More information can be found on our website;

www.creativekiwis.com

or at our blog

http://leftfieldnz.wordpress.com/

http://nzwriter.wordpress.com/

or our RSS Feeds at

http://leftfieldnz.wordpress.com/feed/

http://nzwriter.wordpress.com/feed/

We also have a free ebook available at
http://ubuntuone.com/3kuJ8trer8kf4tr07hyP8F

We hope you enjoyed the experience.

Thank you again.

Bill and Craig.

Two Centuries of Telecommunications in My Lifetime

by Bill Rosoman

Two Centuries of Telecommunications in My Lifetime, is the summary of all the telecommunication changes in my short 64 year life, it is amazing how far we have come in the last 50 odd years.

Keywords
Telecommunication, Phones, Smart Phones, Telex

###

www.ingramcontent.com/pod-product-compliance
Lightning Source LLC
LaVergne TN
LVHW010107110826
845155LV00028B/529

* 9 7 8 1 9 2 7 1 5 7 2 8 2 *